AF810093

SOFTWARE ENGINEERING

SOFTWARE ENGINEERING

Ishaq Azhar Mohammed
Dhaya Sindhu Battina
Dilip Kumar Varma Jetty

red'shine
Publication
INDIA

SOFTWARE ENGINEERING

by: Ishaq Azhar Mohammed, Dhaya Sindhu Battina,
Dilip Kumar Varma Jetty

■

RED'SHINE PUBLICATION PVT. LTD.
Headquarters (India): 88-90 REDMAC, Navamuvada,
Lunawada, India-389 230
Contact: +91 76988 26988
Registration no. GJ31D0000034

In Association with,
RED'MAC INTERNATIONAL PRESS & MEDIA. INC
India | Sweden | UK

■

■

■

ISBN: 978-93-91479-87-9
ISBN-10: 9-39-147987-1
DIP: 18.10.9391479871
DOI: 10.25215/9391479871
Price: ₹ 400
November, 2021 (First Edition)

■

■

www.redshine.co.in | info@redshine.in
Printed in India | Title ID: 9391479871

INDEX

CHAPTER NO.	CHAPTER NAME	PAGE.NO
Chapter-1	1.0 Introduction	01
	2.0 Software Engineering	01
	2.1 History	01
	2.2 Software Process & Its Characteristics	02
	3.0 Phases in Software Development	05
	4.0 Software Development Process Models	11
Chapter-2	1. Introduction: SRS and Needs	20
	2. Problem Analysis: (Structuring Information ONLY)	23
	3. Requirement Specifications: Characteristics & Components of SRS	25
	4. Specification Languages (Structured English, Regular Expressions & Decision Tables)	27
	5. Structure of SRS	28
	6. Validation of SRS	30
	7. Introduction to software projects, Planning, Categories of Software projects	32
	8. Overview of Cost estimation, Uncertainty in cost estimation, size estimation	34
	9. COCOMO Model	36
	10. Project Monitoring Plan (Time sheets, Reviews, Cost-Schedule-Milestone, Earned Value Method)	41
Chapter 3	1. Introduction: System Design, Design Objectives	45
	2. Design Principles / Concepts: Top-down & Bottom-up approach, Problem Partitioning, Abstraction, Modularity	48
	3. Module Level concept, Coupling, Cohesion	53
	4. Overview of structured design	59
	5. Functional v/s Object–Oriented approach (Difference only)	60
	6. Design specification, Verification	60
	7. Introduction: Detailed design, Module specification and its desirable properties, Functional module specification, Data	63

- During 1950, the concept of operating systems started and languages like FORTRAN and COBOL were introduced. Thus now the hardware was not much important, but the programs were important.
- In 1960s multiprogramming OS were found. Thus, prices were of hardware decreased.
- With the availability of cheaper and more powerful machines, higher-level languages, and more user-friendly operating systems, the applications of computers grew rapidly.
- The techniques for writing simple programs could not be scaled up for developing software systems. Now there was need of some standard techniques to write programs. This was called Software crisis.
- There were two meetings held by NATO (North Atlantic Treaty Organization) Science committee and the term "Software Engineering" was coined.
- With time, the cost of hardware decreased, and for software increased. This is because the development of software has many people involved and time taken is also more.

Definition

- **Software Engineering** is systematic approach to development, operation, maintenance and retirement of software.
- Goal of Software Engineering: To produce high-quality software at low cost.

2.2 Software Process & its characteristics

- **Software Process:** A software process specifies a method of developing software.
- **Software Project:** A software project is a development project in which a software process is used.
- **Software Product:** Software product is the outcome of a software project.

INDEX

CHAPTER NO.	CHAPTER NAME	PAGE.NO
Chapter-1	1.0 Introduction	01
	2.0 Software Engineering	01
	2.1 History	01
	2.2 Software Process & Its Characteristics	02
	3.0 Phases in Software Development	05
	4.0 Software Development Process Models	11
Chapter-2	1. Introduction: SRS and Needs	20
	2. Problem Analysis: (Structuring Information ONLY)	23
	3. Requirement Specifications: Characteristics & Components of SRS	25
	4. Specification Languages (Structured English, Regular Expressions & Decision Tables)	27
	5. Structure of SRS	28
	6. Validation of SRS	30
	7. Introduction to software projects, Planning, Categories of Software projects	32
	8. Overview of Cost estimation, Uncertainty in cost estimation, size estimation	34
	9. COCOMO Model	36
	10. Project Monitoring Plan (Time sheets, Reviews, Cost-Schedule-Milestone, Earned Value Method)	41
Chapter 3	1. Introduction: System Design, Design Objectives	45
	2. Design Principles / Concepts: Top-down & Bottom-up approach, Problem Partitioning, Abstraction, Modularity	48
	3. Module Level concept, Coupling, Cohesion	53
	4. Overview of structured design	59
	5. Functional v/s Object–Oriented approach (Difference only)	60
	6. Design specification, Verification	60
	7. Introduction: Detailed design, Module specification and its desirable properties, Functional module specification, Data	63

CHAPTER NO.	CHAPTER NAME	PAGE.NO
	abstraction specification. 8. Verification – Design Walkthrough, Critical Design, Review, Consistency checkers	70
Chapter 4	1. Introduction: Coding, Top Down & Bottom Up Approach for coding 2. Structured Programming, Information Hiding 3. Programming Style 4. Internal Documentation 5. Verification (Code Reading ONLY) 6. Introduction: Testing, Error, Fault, Failure & Reliability 7. Top down and bottom up approach for testing 8. Levels of Testing 9. Functional Testing v/s Structural Testing (Difference ONLY) 10. Automated Testing Tool	73 74 76 78 78 79 81 83 85 85

- During 1950, the concept of operating systems started and languages like FORTRAN and COBOL were introduced. Thus now the hardware was not much important, but the programs were important.
- In 1960s multiprogramming OS were found. Thus, prices were of hardware decreased.
- With the availability of cheaper and more powerful machines, higher-level languages, and more user-friendly operating systems, the applications of computers grew rapidly.
- The techniques for writing simple programs could not be scaled up for developing software systems. Now there was need of some standard techniques to write programs. This was called Software crisis.
- There were two meetings held by NATO (North Atlantic Treaty Organization) Science committee and the term "Software Engineering" was coined.
- With time, the cost of hardware decreased, and for software increased. This is because the development of software has many people involved and time taken is also more.

Definition

- **Software Engineering** is systematic approach to development, operation, maintenance and retirement of software.
- Goal of Software Engineering: To produce high-quality software at low cost.

2.2 Software Process & its characteristics

- **Software Process:** A software process specifies a method of developing software.
- **Software Project:** A software project is a development project in which a software process is used.
- **Software Product:** Software product is the outcome of a software project.

CHAPTER - 1

1.0 INTRODUCTION

SOFTWARE DEFINITION

- Software is the collection of computer programs, procedures, rules and related documentation and data.
- It is used by many people instead of only developer.
- It requires sufficient documentation as help to user.
- Programs are tested perfectly because users are not allowed to make changes.

Difference between Program & software

Program	Software
Small in size (LOC)	Large in size (KLOC)
Limited Functionality	Wide range of functionality
Simple or less complex	Complex
Used by the author of the program	Used by people other than developers
Little or no documentation	Sufficient documentation
Bugs can be tolerated	Bugs can not be tolerated
Testing is often not done.	Testing is always done.
Portability, reliability & usability are not so important.	Portability, reliability & usability are also important.

2.0 SOFTWARE ENGINEERING

2.1 History

- The electronic computers started during 1940s. During this time, there was more concentration on the development of hardware. There were no operating systems and the programs were written on paper tapes or switches.

The fundamental objectives of a process are:
Optimality: Means that the process should be able to produce high quality software at low cost.
Scalability: Means that it should be applicable for large software projects.

To achieve these objectives, a software process should have some characteristics / properties.

Characteristics will describe what features should be available in Software Process to make the software product better quality. There are mainly 4 characteristics of Software Process:

1. Predictability
2. Support Testability & Maintainability
3. Early Defect Removal & Defect Prevention
4. Process improvement

1. Predictability:

It determines how accurately the outcome of a process in a project can be predicted before the project is completed.

For example, the project A is very similar to project B that we did 2 years ago. Hence, A's cost will be very close to B's cost. This implies that the process will be used to develop project A will be the same as the process used for project B. In such a situation, we can say that process is **predictable**.

If process is predictable then we can determine cost and quality of a project. It should be clear that if we want to use past experience to control cost and ensure quality, we must use a predictable process.

2. Support Testability & Maintainability:

Any software may take few years in development and can be used for 5-20 years. The cost of development is the total cost of its phases – requirement analysis, design, coding and testing.

Moreover, there is also the cost of maintenance after software is prepared. And this cost is more than development cost. The ratio of cost of development to maintenance is given as 40/60, 30/70 or even lower.

A typical Efforts Distribution is given as below:

Requirement Analysis	10%
Design	20%
Coding	20%
Testing	50%

Testing part contains major part because it is very important and quality of software depends on it.

The goal of the process should not be to reduce the effort of design and coding, but to reduce the cost of testing and maintenance.

Testing and Maintenance depends on design and coding hence software must be designed and coded to make testing and maintenance easier.

3. Early Defect Removal and Defect Prevention:

Errors can occur at any stage during development.

A typical error distribution is given as:

Requirement Analysis	20%
Design	30%
Coding	50%

The cost of correcting errors of different phases is not same. It depends on when it was detected and corrected.

If the error is detected I the later stages, the cost of correcting will be higher. Cost of correction of the error found in coding stage is very high compared to error found in the requirement stage.

Thus we should attempt to detect errors that occurs in a phase during the phase itself and should not wait until testing to detect errors.

4. Process improvement:

The goal of software engineering is to improve the quality of software and to reduce the development cost of the product. But the quality and cost of the software is determined by the process. So, the software process must be improved.

Process improvement can be done by evaluating the existing process and understanding the weakness in the process.

For quality improvement and cost reduction the software process must be improved.

3.0 PHASES IN SOFTWARE DEVELOPMENT

A software development process consists of various phases, each phase ending with a defined output.

The main reason for having phases is that it ensures that the cost of development is lower compared to if it was developed as a whole.

It allows proper checking for quality and progress at some defined points. Otherwise one will have to wait until entire software is developed completely. Developing software as a whole result in complexity.

is **system design document**. In this part we decide major Data Structures, File formats, Output Formats and main modules.

Detailed Design specifies internal logic of each module. This logic is given in high-level language (and not in a language in which software is developed). The input of a detailed design is system design document. The output of a detailed design is **detailed design document**.

A design methodology is a systematic approach to creating a design by application of set of techniques and guidelines.

3. Coding

Definition: This phase is related to the programming used for the system.

Goal: To translate the design into the code of required programming language in software.

Description:

This phase affects the testing and maintenance. If the code is well written, it can reduce the cost of testing and maintenance.

The **standard used** for coding is **Structured programming**. It means that programs should be written in sequence of statements and statements should be executed in sequence.

To verify this step, we may read the code by review technique. The testing of coding by executing it is called 'unit testing'. Here we run each module of system rather than running the whole system. So this phase is called "Coding and Unit Testing".

The **output** of this phase is unit-tested code of different modules.

4. Testing

Definition: Testing is the measure to check the quality of software during the development.

Goal: To detect the errors in the requirement, design and coding phases.

This phase starts with test plan, which identifies all the activities that must be performed for testing.

During testing, a test case specification document is produced, which lists different test cases, together with expected outputs. The specified test cases are executed and actual result is compared with expected output.

There are 4 types of Testing:

1) Unit Testing
2) Integration Testing
3) System Testing
4) Acceptance Testing

1) Unit Testing: It is a process to test each of the modules. Different modules are tested against the specifications produced during design of the modules. Unit testing is essential for verification of the code produced during the coding phase. Goal of unit testing is to test internal logic of the modules. It is done by the programmer of the module.

2) Integration Testing: Relations between modules are considered. It is a process to test the system after combining all the modules. Many unit-tested modules are combined unto subsystem, which are then tested. The goal of Integration Testing is to see that modules are integrated properly.

3) **System Testing**: System Testing is performed to check that system is working according to the requirement.

4) **Acceptance Testing:** Finally, acceptance testing is done in which client will enter the actual data and check the working of software.

The methods of testing are:

1. **Functional Testing**
2. **Structural Testing**

Functional Testing	Structural Testing
It is also known as **"Black Box Testing"**.	It is also known as **"Glass Box Testing"**.
In this testing, input and output are known but the actual process is unknown.	In this testing, input and output are unknown but the actual process is known.
In functional testing, the internal logic of the system under testing is not considered.	In structural testing, the internal logic of the system under testing is considered.
The test data is given according to requirement because it does not know what is written inside code.	The test data is given to check the logic of code.

The final output of testing phase is **Test Report and the Error report**. Test report contains set of test cases along with actual output. Error report describes the error found and action taken to remove the errors.

Maintenance

Maintenance is not part of software development but it is an important activity in life of software.

Major type of maintenance:

Adaptive Maintenance:

In this type of maintenance there is no error in the system but there may be some changes required by the Client in the input and output formats, or in the requirement analysis.

Corrective Maintenance:

In this case there are errors in software and they are solved.

4.0 SOFTWARE DEVELOPMENT PROCESS MODELS

Software Development Process is a set of activities and when they are performed in order, it produces the desired software. For example, design, coding and testing.

Process Model: Process Model specifies some activities that should be performed and the order in which they should be performed.

WATERFALL MODEL

It is the simplest process model. It is most widely used model. The phases in this model are arranged in linear order.

The sequence of activities in waterfall model is:

Requirement Analysis, Project Planning, System Design, Detailed Design, Coding, Unit Testing, System Integration and Testing followed by Installation, Operations and Maintenance.

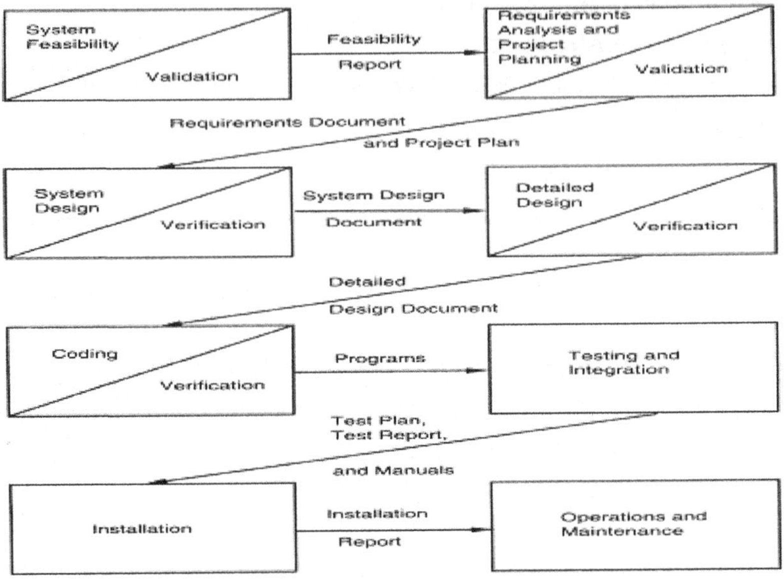

The **waterfall model** is using linear order because of the following two **assumptions**:

1. For successful project (resulting in successful product), all the phases listed in waterfall model must be performed anyway.
2. Any different order of phases will result in less successful software product.

In this model, a project begins with feasibility analysis. If the project is found feasible then further activities are carried out.

At the end of each phase, verification or validation is performed to ensure that output of each phase is consistent with its input and with overall requirements. There is a defined output after each phase which is certified before the next phase begins.

To certify an output product of a phase before next phase begins, reviews are often held. Reviews are formal meetings to uncover deficiencies in a product. The review reports are the outcomes of these reviews.

Limitations of Waterfall Model:

1. **The waterfall model assumes that requirements can be frozen i.e. fixed before the design begins.**
   This is possible for systems which are designed to automate an existing manual system. But for new systems, determining requirements is difficult, as the user does not even know the requirements.

2. **Due to freezing requirements, usually hardware specification is also fixed.**
   A large software project may take few years to complete. It may happen that by time hardware technology may become obsolete because hardware technology is changing fast.

3. **This model suggests that the requirement must be completely specified before the development.**
   Sometimes it is desirable to first develop a part of system completely and then later enhance the systems in phases.

4. **This is not suitable for interactive application.**
   This model is requiring formal documents at the end of each phase. This approach is not suitable for interactive applications, where developing detailed documentation of user interfaces is not feasible.

It is **well suited** for routine types of projects, where requirements are well understood. That is waterfall model is suitable if we are familiar with the problem domain and requirements are clear.

PROTOTYPING MODEL

The goal of this model is to overcome the first two limitations of the waterfall model.

In this model requirements are not freezed, but a prototype is created as per the current requirements. The development of prototype starts when preliminary version of the requirements specification has been developed.

After the prototype has been developed, the end users and clients use the prototype.

Based on their experience, they provide feedback to the developers. Based on the feedback the prototype is modified and the clients are again allowed to use the system.

The cycle repeats until the users and clients are completely satisfied. The initial requirements specification is modified to final requirement specification.

Development of prototype obviously undergoes design, coding and testing, but each of these phases not done very formally and thoroughly.

The concept of the model is shown as below:

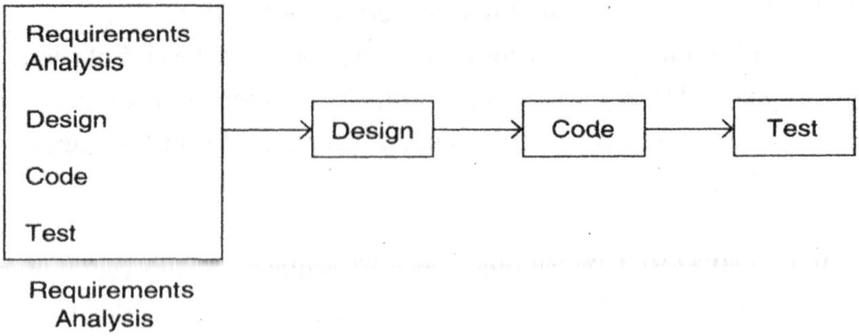

Advantage:

1. By this model the client gets the actual feel of the system, because this enables the client to better understand the requirements.

2. This **model is useful** for the large and complicated system, which is totally new, and there is no existing system or manual process for it.
3. In some situations, the cost of software development without prototyping may be more than with prototyping. Even if it is built twice, the overall cost of development is less, as this model is precise.
4. The cost of testing and documentation is reduced.
5. The experience of the development is useful to the developers, and thus developing the similar projects of experience will have the less cost.

Use:

Prototype model is well suited for projects where requirements are hard to determine and the confidence in obtaining the requirements is low.

In such projects requirements may change and rework is required.

Prototype model is generally not used for following reasons: (Disadvantage)

1. The cost of development is more, because it is similar to building software twice.
2. It is not complete and many details are missing.
3. Exception handling, Recovering and implementation of standards are not included here.

ITERATIVE ENHANCEMENT MODEL

This model is to remove the third limitation of the waterfall model and combines the benefits of both prototyping model and waterfall model.

According to this model, the software is developed in terms of the increments, and each increment adds some functionality in the system. This process is continued until the final system is obtained.

First a version is released that contains some capability. Based on feedback from users and experience from users and experience from this version a list of additional desirable features and capabilities are generated. These features forms the basis of the software, and are included in next version

This model is represented as below:

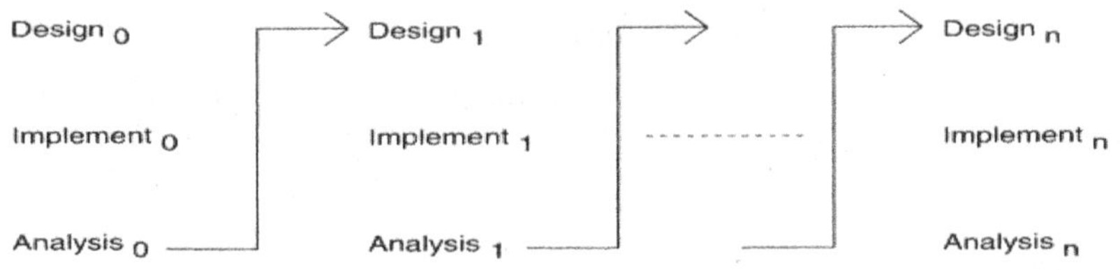

The phases in this system are Design, Implementation and analysis.

In the first step, the simple initial implementation is done for overall system. Project Control list is prepared, which gives the list of all tasks that are performed.

The process continued until the project control list becomes empty.

The advantage of using control list is that it minimizes the chance of error, and reduces the redesigning.

Use:

1. This model is **applicable for product development**, in which the developers themselves provide specifications. Therefore, have a lot of control on what specifications go in the system and what stay out.
2. It is also **applicable in client-oriented projects.** Client's organization does not have to pay for the entire software together. They can get main part of the software developed and

perform cost-benefit analysis for it before enhancing the software with more capabilities.

Advantage:

1. This process model can be useful if the core of the application to be developed is well understood and the increments can be easily defined.
2. This model gives better testing, because there is testing of each phase like waterfall model. The requirements of system are provided as feedback like prototyping.

SPIRAL MODEL

It's a **risk-driven model.** This model is very useful in high risk project. It is the recent model proposed by Boehm. The activities of spiral model are arranged in the spiral manner.

There are two dimensions in spiral model.

1. Radial Dimension
2. Angular Dimension

The radial dimension represents the cumulative cost occurred in each step, and angular dimension represents the progress made in completing each cycle.

CHAPTER - 2

1. SOFTWARE REQUIREMENT SPECIFICATION (SRS)

Software Requirement Specification (SRS) is a document that completely describes what the proposed software should do without describing how the software will do it. It is the starting point of the software development activity. The developers could clearly understand the system when it was explained to them. But when systems became more complex, this phase becomes more important. For the large systems now, Requirments Specification is most difficult and error prone.

In this phase, the requirement analyst (developer) has to identify the needs of the system by taking to the clients (users). If some requirements are added into the existing system, then it is called "new features" of the software.

"SRS is a process of converting ideas in minds of clients (input), into the formal document (output)."

1.1 Need / Role (Benefits) of SRS

In the development of software there are three parties involved: the client, the users and the developer. In the development, the ideas required by the client must be transformed into the software, by the developer. But the problem is that the client does not know software and developer does not know the application area (business). Thus, there is a communication gap between them.

SRS is therefore used to bridge this gap. A good SRS should satisfy all parties, which is very hard. Goals of a good SRS are as follows:

perform cost-benefit analysis for it before enhancing the software with more capabilities.

Advantage:

1. This process model can be useful if the core of the application to be developed is well understood and the increments can be easily defined.
2. This model gives better testing, because there is testing of each phase like waterfall model. The requirements of system are provided as feedback like prototyping.

SPIRAL MODEL

It's a **risk-driven model.** This model is very useful in high risk project. It is the recent model proposed by Boehm. The activities of spiral model are arranged in the spiral manner.

There are two dimensions in spiral model.

1. Radial Dimension
2. Angular Dimension

The radial dimension represents the cumulative cost occurred in each step, and angular dimension represents the progress made in completing each cycle.

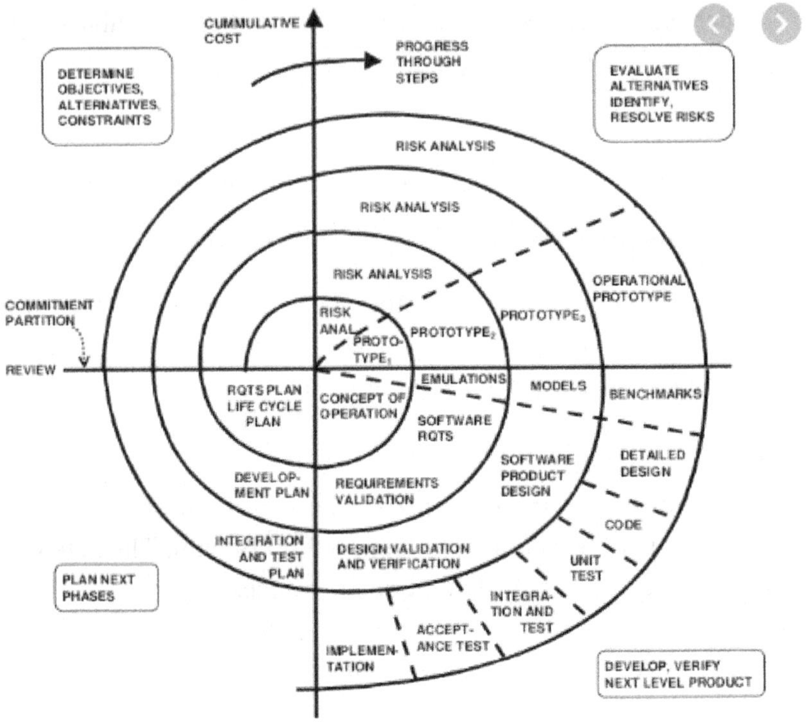

Each phase of Spiral Model is divided into **four quadrants** as shown in the above figure. The functions of these four quadrants are:

1. Identify the objectives at the start of every phase. Then possible alternative solutions are proposed.
2. Evaluate all the possible solutions to select the best possible solution. Then the risks associated with that solution is identified. And the risks are resolved using the best possible strategy. At the end, Prototype is built for the best possible solution.
3. Identified features are developed and verified through testing. At the end, the next version of the software is available.
4. Customers evaluate the so far developed version of the software. In the end, planning for the next phase is started.

For a typical project using this model, it starts with the **round zero of feasibility study**, and continue with round one, and so on until the actual development is over.

Advantage:

1. Each cycle in this model is completed by the review of the product as well as the plans. Thus it is useful for Development (new) as well as Enhancement (existing to new) projects.
2. Most applicable with high risk projects.
3. The development progress and cost can be identified at any stage.
4. The risk management provides secure development.

Disadvantage:

Spiral Model is not so well-known as other SDLC (Software Development Life Cycle) models because

- It can be pretty costly to use and
- Doesn't work well for small projects.

CHAPTER - 2

1. SOFTWARE REQUIREMENT SPECIFICATION (SRS)

Software Requirement Specification (SRS) is a document that completely describes what the proposed software should do without describing how the software will do it. It is the starting point of the software development activity. The developers could clearly understand the system when it was explained to them. But when systems became more complex, this phase becomes more important. For the large systems now, Requirments Specification is most difficult and error prone.

In this phase, the requirement analyst (developer) has to identify the needs of the system by taking to the clients (users). If some requirements are added into the existing system, then it is called "new features" of the software.

"SRS is a process of converting ideas in minds of clients (input), into the formal document (output)."

1.1 Need / Role (Benefits) of SRS

In the development of software there are three parties involved: the client, the users and the developer. In the development, the ideas required by the client must be transformed into the software, by the developer. But the problem is that the client does not know software and developer does not know the application area (business). Thus, there is a communication gap between them.

SRS is therefore used to bridge this gap. A good SRS should satisfy all parties, which is very hard. Goals of a good SRS are as follows:

1. Create the basis for agreement between client and developer, about software.

Justification:

There are three major parties involved in development of new system: the client, the user and developer. The client usually does not understand the software and the software development and the developer do not understand the client's problem and application area. This causes communication gap between the parties involved in the development project. The basic purpose of SRS is to bridge this communication gap.

SRS is a medium through which client and user needs are accurately specified.

SRS forms the basis for software development. A good SRS should satisfy all the parties. Through SRS client clearly describes what it expects from supplier and the supplier clearly understands what capabilities to build in the software.

Thus client and supplier agree on what is required and what is to be developed through an SRS.

2. Reducing the development cost.

If SRS contains the perfect information, there is very less chance of problems in final software. Thus, there is reduction in cost.

Justification:

The quality of SRS has an impact on cost of project.
Errors can exist in SRS.
The cost of fixing the errors increases almost exponentially as time progresses.

If an error in requirement phase is detected and removed after system has been developed, it can cost up to 100 times more than removing it during the requirements phase itself.

From the relative cost of fixing errors as reported in a multi-company study, the approximate average cost of fixing requirements errors depending on phase is as shown below:

Phase	Cost (persons-hours)
Requirements Analysis	2
Design	5
Coding	15
Acceptance Test	50
Operations &Maintenance	150

As shown in table the requirement analysis phase involves cost of engaging 2 persons for error fixing and this cost increases subsequently as we move on to next phase and it is maximum during Operations and Maintenance phase.

There is tremendous reduction in project cost if errors in SRS solved during requirement phase only.

3. Providing the references for validation of software. When the software is prepared, we can check (Validate) whether it is according to requirement.

Justification:

An SRS is document that specifies the requirements of users and client.

An SRS helps client to determine if software meets the requirements specified by the client and the user.

When software is developed completely we need to determine if the software being delivered is what was ordered.

Without proper SRS, the client cannot determine whether software is developed as per the requirements. Even the supplier cannot convince the client that all requirements are fulfilled.

SRS acts as proof for users and clients to validate the product against the requirements.

4. A high-quality SRS is prerequisite to high-quality software.
Justification:
For a high-quality SRS a high-quality SRS is a must.

Many errors are made during requirement analysis phase. An error in SRS is most likely to appear in the final system.

If SRS document specifies the wrong system then even correct implementation of SRS will lead to a system that will not satisfy the client.

Thus if we want a high-quality software we must begin with high-quality SRS.

2. PROBLEM ANALYSIS

This is the first step involved in requirement specification. The aim of problem analysis is to obtain clear understanding of the needs of the clients and users, what a software should do and what are the constraints that must be considered.

Issues involved in Analysis:

1. To obtain the necessary information (sources and methods)
2. To organize the information obtained (completeness and consistency)
3. To resolve contradictions (consistent final specifications)
4. To avoid internal design (focus on what)

There are certain activities involved in this phase as below:

- To collect information regarding existing system and requirements of users and clients.
- To organize the information so that it can be evaluated for completeness and consistency.
- To resolve the contradictions that may exist in the information from different parties.

2.1 Structuring Information

During the analysis, getting the information about system is not difficult, but which information should be collected is difficult. There are three principals for structuring the information: **Divide and Conquer (Partitioning)**, **Abstraction** and **Projection**.

Divide and Conquer / Partitioning

In this technique the analyst divides the problem into sub problem. And try to understand each sub problem and its relationship to other sub problem to understand the total problem. For example, the operating system cannot be explained directly, but we can explain all its parts like files, memory management etc.

This partitioning is done with respect to:

1. Objects
2. Functions

An object is an entity in real world that has clearly defined boundaries and an independent existence.

Function is a task, service, process or a mathematical function, or an activity that is now being performed in the real world or has to be performed by the system that will be built to solve to solve real world problem.

Abstraction: In abstraction, the system is defined in general terms, and then the details are provided separately. For example, when file system of Operating System is analyzed, we provide the details of different functions in file system. Here when we want to study the details of one part of the system, there is no need to understand other parts.

Projection: This is used to many times for structuring information. It is used to define the system with multiple viewpoints. For e.g. seeing the 3D object form different angles like top and side. Here the system is studied with different projections, like Operating System can be analyzed as per programmer, general user and expert. Advantage of projection is that problem of difficulty and error prone is reduced as we see all aspects of the system.

3. REQUIREMENT SPECIFICATION

3.1 Characteristics of SRS

1) **Understandable:** The SRS document must be such that the developer, client and the users can understand.
2) **Unambiguous (unmistakable):** SRS can be unambiguous only if all the requirement has one and only one interpretation (meaning). Since it is written in natural language, there are possibilities of ambiguity (confusing words). Some formal language must be used to remove it.
3) **Complete:** SRS is complete if everything of software is specified. It represents the effect of system for all types of data.
4) **Verifiable:** SRS is verifiable, if we can check it to compare with the final software.
5) **Consistent (reliable):** SRS is consistent, if there are no requirements which are conflicting (opposing) each other. For e.g. in one process the event A occurs only after B, in some other process, event B occurs only after A. this type of inconsistency is removed.
6) **Modifiable:** Sometimes we may want to make some changes or add some new features in system. Thus, SRS must be modifiable.

7) **Traceable:** It means that SRS is clearly converted to software. Forward traceability means requirement must be converted to design & code. Backward traceability means design & code can be traced to requirement.

3.2 Components of SRS

1) Functional Requirement:

It specifies that what is the output for a given input. It shows the relation between input and output of the system. In this step it is very important to specify the behavior of system abnormal situations. For e.g. it should give solution to what will happen if invalid data is entered. Also, we should specify that even if the input is valid, which operations are not possible.

2) Performance Requirement:

There are two types of performance requirements: **Static** & **Dynamic**.

Static requirement is fixed (which do not change). For e.g. Total no. of terminals supported, no. of users and no. of files and its sizes. It is also called capacity requirement.

Dynamic requirement means those which are related to execution of the system. It includes two main aspects: Response time & Throughput.

3) Design Constraints:

It represents the restrictions (limitations) which are present when the developer is designing the system. There are following four constraints:

a) **Standards compliance:** It specifies all the standards which system must follow. The standard may be report format and accounting procedures.

b) **Hardware Limitations:** The software must operate only on the hardware which is available to us. There may be restrictions in type of machine, type of O.S., languages and the limits of primary & secondary storage.

c) **Reliability & Fault tolerance:** Reliability requirement is useful for very complex systems. Fault tolerance specifies the behavior of system when some fault occurs in it.

d) **Security:** It is useful in defense system & many database systems. We can create security by limiting use of commands, giving access rights, use of passwords and cryptography techniques.

4) External Interface Requirement:

It specifies all types of interface with the system, which are **people**, **hardware** and other **software**.

User Interface is very important and proper attention should be given for it. We should clearly specify requirement instead of "the system should be user friendly", we must write perfectly that "system should not have more than 6 char command".

For **hardware**, we must specify that which type of hardware is used.

For **other Software**, we must specify whether out system interface with operating system and take help of other application (system) or not.

4. SPECIFICATION LANGUAGES

Languages used to write SRS are called Specification Languages. The language should process characteristics of SRS. It should be easy to learn and use.

4.1 Structured English

Natural Languages have been widely used for specifying requirements. The major advantage of using natural languages is that both client and supplier understand the language. For small software system the requirements conveyed verbally but in complex (large) software system were specified in written form using natural language. But requirements written in natural language will be imprecise and ambiguous. To overcome these drawbacks, we make use of formal language. Choosing English as natural language and using it in structured fashion is called Structured English.

4.2 Regular Expressions

Regular expressions can be used to specify structure of symbol strings. String specification is useful for specifying such things as input data, command sequence, and contents of message. Regular expressions can be considered as grammar for specifying the valid sequences in a language and can be automatically processed. There are the basic constructs allowed in regular expressions: Atoms, Composition, Alternation, and Closure.

4.3 Decision Table

Decision table provide a mechanism for specifying decision logic. It is formal, table – based notation that can also be automatically processed to check for qualities like completeness and lack of ambiguity. A decision table has two parts. The top part lists the different conditions, and the bottom part specifies different actions.

5. STRUCTURE OF SRS

All the requirements for the system have to be included in a document that is clear and concise. For this, it is necessary to organize the requirements document as sections and subsections. There can be many ways to structure a requirements document.

Many methods and standards have been proposed for organizing an SRS. One of the main ideas of standardizing a structure of the document is that with an available standard, each SRS will fit a certain pattern, which will make it easier for others to understand. The general structure of an SRS is given below:

1. Introduction
1.1. Purpose
1.2. Scope
1.3. Definitions, Acronyms and abbreviations
1.4. References
1.5. Overview

2. General Description
2.1. Product Perspective
2.2. Product Functions
2.3. User Characteristics
2.4. General Constraints
2.5. Assumptions & dependencies

3. Specific Requirements
3.1. External Interface Requirements
    3.1.1. User Interfaces
    3.1.2. Hardware Interfaces
    3.1.3. Software Interfaces
    3.1.4. Communication Interfaces
3.2. Functional Requirements
    3.2.1. Mode 1
        3.2.1.1. Functional Requirement 1.1
            3.2.1.n Functional Requirement 1.n
            3.2.m Mode m
            3.2.m.1 Functional Requirement m.1
            3.2.m.n Functional Requirement m.n

3.3. Performance Requirements
3.4. Design Constraints
3.5. Attributes
3.6. Other Requirements

First (introduction) section contains the purpose, scope, overview, etc. of the requirements document. It also contains the references mention in the document and any definitions are used.

Section 2 describes the general factors that affect the product and its requirements. Specific requirements are not mentioned, but a general overview is presented to make the understanding of the specific requirements easily.

The specific requirements section **(section 3)** describes all the details that the software developer needs to know for designing and developing the system. This is typically the largest and most important part of the document.

6. VALIDATION OF SRS

The development of software starts with the requirements document, which is also used to determine whether delivered software is the one which was ordered. That is whether the software satisfies all the requirements stated in SRS. So, it is necessary that SRS does not have errors and it should clearly specify the client's requirements correctly.

The basic objective of the requirements validation activity is to ensure that the SRS reflects the actual requirements accurately and clearly. A related objective is to check that the SRS document is itself "good quality". Many types of errors are possible, but the most common errors that occur in the SRS are omission, inconsistency, incorrect fact and ambiguity.

There four types of errors that may occur in SRS:

1) Omission: When a requirement is not included in SRS then such a error is called Omission.

2) Inconsistency: When a requirement is contradictory to some other requirement or stated requirements are not compatible with actual requirements then such error is called inconsistency.

3) Incorrect Fact: When a fact recorded in SRS is incorrect then such an error is called Incorrect fact.
4) Ambiguity: When requirement stated in SRS have multiple meanings and its interpretation is not unique then such an error is called Ambiguity.

6.1 Validation Methods

1) **Automated Cross Referencing:** It is the used to check the document automatically that there are some standard languages like PSL (Problem Statement Languages) and SADT (Structured Analysis and Design Techniques). We can compare document with standard and remove errors if any.
2) **Requirement Review:** It is the most common method used for validation. The review of requirement is done by a group of people. This group consists of author of SRS document, person who understands needs of clients, person of design team, person responsible of maintaining document & also some persons who are indirectly involved in system.

   By this method, we can review incompleteness, inconsistency, unfeasibility and incorrect translation.

   - **Incompleteness** means that there may be some functions or modules which are not described completely in detail.
   - **Inconsistency** means that there is contradiction (opposing statement) within document like in starting it is written that code must be less than 5 char., and afterwards it says that code must have 6 chars.
   - **Unfeasibility** means that requirement is not feasible.
   - **Incorrect translation** means that when document is prepared, the information of system is written incorrectly, may be due to some misunderstanding. Checklist is used for this method, which is list of questions which are confusing about system.

3) **Reading:** The goal in reading is to have someone other than the author of the requirements read the requirements specification document to identify potential problems. By having the requirements read by another person who may have different "interpretation" of requirements, then many of the requirements problems caused by misinterpretations or ambiguities can be identified. Reading is effective only if the reader takes the job seriously and reads the requirements carefully.
4) **Constructing Scenarios:** Scenarios describe different situations of how system will work once it is operational. The most common area for constructing scenarios is that of system-user interaction. Constructing scenarios is good for clarifying misunderstandings in human-computer area. They are of limited value for verifying the consistency and completeness of requirements.
5) **Prototyping:** In this method a prototype is built to verify requirements. Prototypes can be quite useful in verifying the feasibility of some of the requirements. A prototype that has been built during problem analysis can also aid validation. For example, if the prototype has most of the user interfaces and they have been approved after use by the client and users, then the user interface, as specified by the prototype, can be considered validated. No further validation need be performed for them.

7. INTRODUCTION TO SOFTWARE PROJECT AND PLANNING

Software development is a highly labor-intensive activity. A large software project may involve hundreds of people and span many years. A project of this dimension can easily turn into chaos if proper management controls are not imposed. To complete the project successfully, the large workforce has to be properly organized so that the entire workforce is contributing effectively and efficiently the project. Proper management controls and checkpoints are required for effective project monitoring.

For a successful project, expert management and technical staff are both essential. Lack of any one can cause a project to fail. Project management activities can be viewed as having three major phases: Project planning, Project monitoring and control, and Project termination.

In planning, all activities that must be performed before starting the development work. Once the project is started, project control begins. In other words, during planning all the activities that management needs to perform are planned, while project control the plan is executed and updated.

Without a proper plan no real monitoring and controlling of the project is possible. One of the reasons for improper planning is the old thinking that the major activity in a software project is designing and writing code.

The basic goal of planning is to look into the future, identify the activities that need to be done to complete the project successfully, and plan the scheduling and resource allocation for these activities. So, all future activities should be planned.

All economic, political, and personnel factors should be taken into account for a realistic plan and thus for a successful project. The input to the planning activity is the requirements specification. A very detailed requirements document is not essential for planning, but for a good plan all the important requirements must be known. The output of this phase is project plan, which is a document describing the different aspects of the plan.

The major issues the project plan addresses are:

- Cost Estimation
- Schedule and Milestones
- Personnel Plan

- Software Quality Assurance Plans
- Configuration Management Plans
- Project Monitoring Plans
- Risk Management

8. COST ESTIMATION

Cost in a project is due to the requirements for software, hardware, and human resources. Hardware resources are such things as the computer time, terminal time, and memory required for the project, whereas software resources include the tools and compiler needed during development. The bulk of the cost of software development is due to the human resources needed.

Cost estimation is a set of techniques and procedures used to arrive at a cost estimate. These techniques are utilized by the process of cost estimation to compute the output from the given set of inputs. The inputs to the process of cost estimation are also called as cost drivers and the outputs are expressed in the form of efforts, duration, loading, or modified requirements to name a few.

Most cost estimates are determined in terms of person-months (PM). Estimates can be based on subjective opinion of some person or determined through the use of models. It is generally accepted that it is important to have a more scientific approach to estimation through the use of cost estimation.

8.1 Uncertainty in Cost Estimation

One can perform cost estimation at any point in the software life cycle. As the cost of the project depends on the nature and characteristics of the project, at any point, the accuracy of the estimate will depend on the amount of reliable information.

When the product is delivered, the cost can be accurately determined, as all the data about the project and the resources spent can be fully known. When the project is being initiated or during feasibility study, at this time we have only some idea of the types of data the system will get and the major functionality of the system. So, the cost estimation based on this type of information cannot be accurate.

Once the requirements are completely specified, more accurate cost estimates can be made compared to the estimates after the feasibility study. Once the design is complete, the estimates can be made still more accurately. As we specify the system more fully and accurately, the uncertainties are reduced and more accurate cost estimates cane be made. The obtainable accuracy of the estimates as it varies with the different phases is shown in figure.

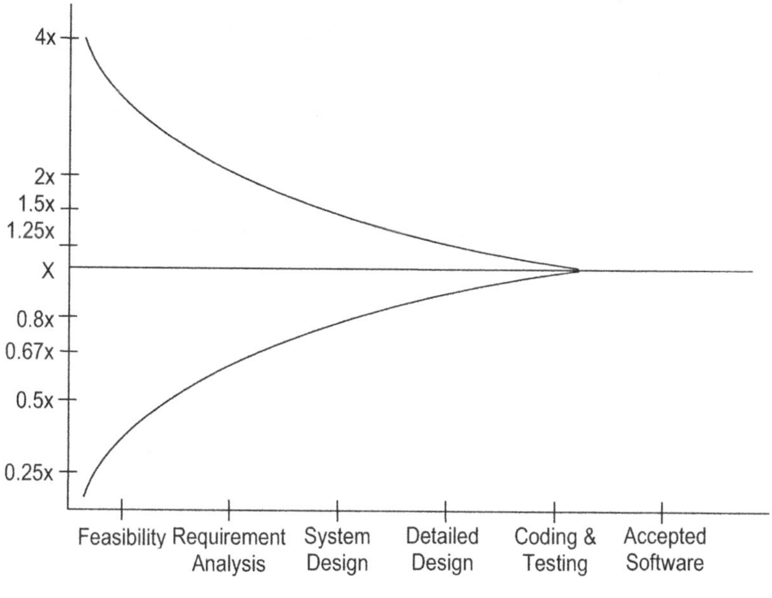

For actual cost estimation, cost estimation models or procedures are developed. The accuracy of the actual cost estimates will depend on the effectiveness and accuracy of the cost estimation procedures or models used.

8.2 Size Estimation

Size of software is the most important part for estimation of cost. If there is inaccuracy in the size, then the cost will also be inaccurate. There were few solutions provided to find the size of a project. However, none of these is a standard & accurate method.

- **Weinberg-Schulman's Experiment:** According to this experiment the programmer should be able to satisfy the goal given to them. Many experiments were made on the programs and it's was successful.
- **Underestimation of software size:** This is the optimistic approach used with a desire to please others.
- **Experience:** The estimate of a size can be based on previous experience with the similar project. However, people may not remember the experiences correctly.
- **Get Details:** The programmers can get the details of each & every phase and thus can approach to more precise value of size.

9. Constructive Cost Model (COCOMO Model)

Instead of having resource estimates as a function of one variable, resource estimates can depend on many different factors, giving increase to multivariable models.

One approach for building multivariable models is to start with an initial estimate determined by using the static single-variable model equations, which depends on size, and then adjusting the estimates based on other variables.

This approach implies that size is the primary factor for cost; other factors have a lesser effect. COnstructive COst MOdel is such type which estimates the total effort in terms of person-months of the technical project staff. The effort estimate includes development, management, and support tasks but does not include secretarial and other staff.

The basic steps in this model are:

1. Obtain an initial estimate of the development effort from the estimate of thousands of delivered lines of code (KDLOC).
2. Determine a set of 15 multiplying factors from different attributes of the project.
3. Adjust the effort estimate by multiplying the initial estimate with all the multiplying factors.

To determine the initial effort (also called nominal estimate) E_i in person-months the equation is:

$$E_i = a * (KDLOC)^b.$$

The value of the constants 'a' and 'b' depend on the project type. In COCOMO projects are categorized into three types: **organic, semidetached, and embedded**.

Organic projects are in an area in which the organization has great experience and requirements are flexible. Such systems are usually developed by a small team. Examples of this type of project are simple business systems, simple inventory management system, and data processing systems.

Embedded projects are in area in which organization has little experience and requirements are inflexible for such aspects are interfacing and reliability. Projects of the embedded type are ambitious and novel. Examples are embedded avionics systems and real-time command systems.

Semidetached systems fall between these two types. Examples of semidetached systems include developing a new operating system (OS), a database management system (DBMS), and a complex inventory management system.

The constants 'a' and 'b' for different systems are given in following table:

System	a	b
Organic	3.2	1.05
Semidetached	3.0	1.12
Embedded	2.8	1.20

There are 15 different attributes, called **cost driver attributes**, which determine the multiplying factors. These factors depend on product, computer, personnel, and technology attributes (called project attributes). Below table shows the effort multipliers for different cost drivers.

Product Attributes:

RELY, required reliability
DATA, database size
CPLX, product complexity

Computer Attributes:

TIME, execution time constraint
STOR, main storage constraint
VITR, virtual machine volatility (instability)
TURN, computer turnaround time

Personnel Attributes:

ACAP, analyst capability
AEXP, application experience
PCAP, programmer capability
VEXP, virtual machine experience
LEXP, programming language experience

Project Attributes:

MODP, modern programming practices
TOOL, use of Software tools
SCHED, development schedule

Each cost driver has a rating scale, for each rating, a multiplying factor is provided.

For example, for the product attribute RELY, the **rating scale is very low, low, nominal, high, and very high**. The multiplying factors for these ratings are .75, .88, 1.00, 1.15, and 1.40, respectively.

The multiplying factors for all 15 cost drivers are multiplied to get the effort adjustment factor (EAF). The final effort estimate, E, is obtained by multiplying the initial estimate by the EAF.

$$E = EAF * E_i.$$

By this method, the overall cost of the project can be estimated. For planning and monitoring purposes, estimates required for the different phases is also desirable. In COCOMO, the effort required is considered a defined percentage of the overall effort. The percentage of total effort spent in a phase varies with the type and size of the project.

The percentages for an organic software project are given in below table:

	Size			
Phase	**Small**	**Intermediate**	**Medium**	**Large**
	2 KDLOC	**8 KDLOC**	**32 KDLOC**	**128 KDLOC**
Product design	16	16	16	16
Detailed design	26	25	24	23
Code & unit test	42	40	38	36
Integration & test	16	19	22	25

In COCOMO the detailed design and code and unit testing are sometimes combined into one phase called the programming phase. **COCOMO provides three levels of models of increasing complexity:**

1. Basic
2. Intermediate
3. Detailed

The model discussed is intermediate COCOMO model. The detailed model is the most complex.

9.1 Example of COCOMO Model

Suppose a system for office automation must be designed. The system is divided in to four major modules: data entry, data update, query, and report generator. This project falls in the organic category.

The sizes for the different modules and the overall system were estimated to be:

Data Entry	0.6 KDLOC
Data Update	0.6 KDLOC
Query	0.8 KDLOC
Reports	1.0 KDLOC
TOTAL	3.0 KDLOC

From the requirements, the ratings, along with their multiplying factors are:

Complexity	High	1.15
Storage	High	1.06
Experience	Low	1.13
Programmer Capability	Low	1.17

The initial effort estimate for the project is:

$$E_i = a * (KDLOC)^b$$
$$= 3.2 * (3.0)^{1.05}$$
$$= 10.14 \text{ PM}$$

All other factors had a nominal rating. So, from these the effort adjustment factor (EAF) is:

$$EAF = 1.15 * 1.06 * 1.13 * 1.17 = 1.61$$

Using the EAF, the adjusted effort estimate is:

$$E = EAF * E_i = 1.61 * 10.14 = 16.3 \text{ PM}$$

The office automation system's size estimate is 3 KDLOC, so we will have to use interpolation to get the appropriate percentage. The percentages for the different phases are:

- design – 16%,
- detailed design – 25.83%,
- code and unit test – 41.66%, and
- integration and testing – 16.5%.

With these, the effort estimates for the different phases are:

System Design :	0.16 * 16.3	= 2.6 PM
Detailed Design:	0.258 * 16.3	= 4.2 PM
Code & Unit Test:	0.4166 * 16.3	= 6.8 PM
Integration :	0.165 * 16.3	= 2.7 PM

These effort estimates will be used later during personnel planning.

10. PROJECT MONITORING PLAN

The good plan is useless if it is not planned properly. Whenever the plans are made for a large project, it may require modification, which should also be allowed. There are different methods used for monitoring of project, as below:

Time Sheets

The management has to track the progress of the project and the expenditure incurred on the project. The most common method of keeping track of expenditures is by the use of a time sheet. The time sheet records how much time different project members are spending on the different identified activities in the project.

The time sheet is a common mechanism for collecting raw data. Time sheets can be filled daily or weekly. The data obtained from the time sheets can be used to obtain information regarding the overall expenditure and its breakup among different tasks and different phases at any given time.

For example, the effort distribution with phases in a particular organization can be obtained from the time sheets. By assigning codes to projects, tasks within a project, and the nature of the work being done, time sheet processing can easily be automated.

Reviews

One of the purposes of reviews is to provide information for project control. Reviews provide a definite and clearly defined milestone. It forces the author of a product to complete the product before the review. Having this goal gives some force and motivation to complete the product.

For monitoring, reviews can provide a wealth of information. First, the review schedules can be used to determine how the project is progressing compared to its planned schedule. Then the review reports indicate in which parts of corrective action, such as replacing a junior person with a senior person, can be taken. In addition, review reports provide insight into the quality of the software being produced and the types of errors being detected.

Cost – Schedule – Milestone Graph

A cost – schedule – milestone graph represents the planned cost of different milestones. It also shows the actual cost of achieving the milestones gained so far. By having both planned cost versus milestones and the actual cost versus milestones on the same graph, the progress of the project can be grasped easily.

The x-axis of this graph is time, where the months in the project schedule are marked. The y-axis represents the cost, in dollars of PMs. Two curves are drawn. One curve is the planned cost and planned schedule, in which each important milestone of the project is marked. This curve can be completed after the project plan is made. The second curve represents the actual cost and actual schedule, and the actual achievement of the milestones is marked.

Thus, for each milestone the point representing the time when the milestone is actually achieved and the actual cost of achieving it are marked.

For a hypothetical project whose cost is estimated to be $ 100K. Different milestones have been identified and a curve is drawn with these milestones. The milestones in this project are PDR (Preliminary Design Review), CDR (Critical Design Review), Module 1 completion, Module 2 completion, integration testing, and acceptance testing.

For each of these milestones some budget has been allocated based on the estimates. The planned budget is shown by a dotted line. The actual expenditure is shown with a bold line.

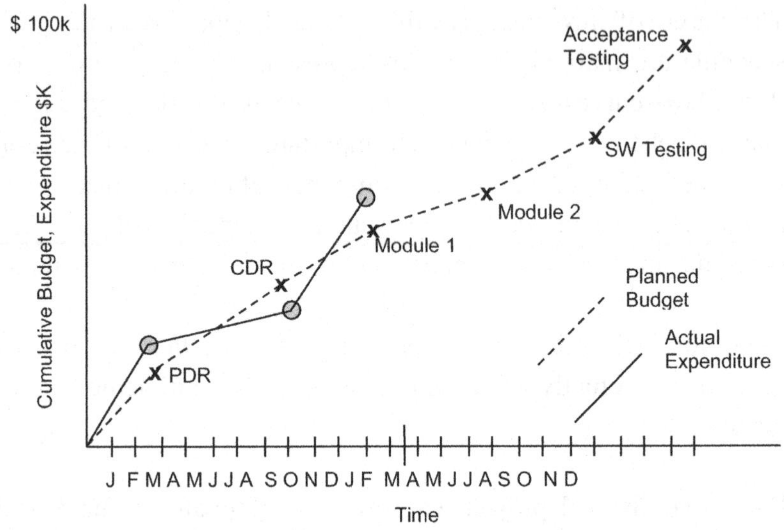

This chart shows that only two milestones have been achieved, PDR and CDR, and though the project was within budget when PDR was complete, it is now slightly over budget.

Earned Value method

It is effective method for project monitoring. It uses STPS (Summery Task Planning Sheet), which is actually a Gantt chart of all the tasks. Earned value means the value which we have assume after the completion of a task. It is in terms of Dollars / PM.

By checking the earned value with the actual cost, we can monitor each part of project. At last we can know the overall cost by adding the cost of all the parts. In general, the value of earned value is somewhat less than the actual effort. At the end of this method, the earned value summery report is produced monthly or weekly.

CHAPTER - 3

1.0 SYSTEM DESIGN

The **design** of a system is a blueprint or a plan for a solution for the system. Design activity begins when requirements document for the software to be developed is available. Requirements specifications activity is a problem domain; design activity is the first step in moving towards solution domain.

Design (as a verb): The process of creating a blueprint or a plan for the system.

Design (as a noun): A blueprint or a plan for the system. It is the output of the design process.

The **goal** of design process is to produce a model or a representation of system, which can be used later to build that system.
The design process has two levels:

- System design
- Detailed design
- A **system design** is first step of a design phase.

At first level the focus is on deciding which modules are needed, specifications of modules, how modules should be interconnected. This is called **system design or top-level design**.

It specifies all the modules of a system and their relationships between them. The input of a system design is SRS documents. The output of a system design is system design document.

During this phase, all the major data structures, file formats, output formats and the major modules in the system and their specification are decided.

- A **detailed design** is second step of a design phase.

In second level, how the specifications are to be satisfied is decided. It is called **detailed design or logic design**.

It specifies internal logic of each module of a system. The input of a detailed design is system design document. The output of a detailed design is detailed design document.

During this phase, details of data structures and algorithm design of each module is specified. The logic of a module is specified in high-level language. It is also known as logic design.

1.1 DESIGN OBJECTIVES

Some desirable objectives for a system design are:

1. Correct
2. Simple and Understandable
3. Complete
4. Verifiable
5. Consistent
6. Efficiency
7. Traceable

1. **Correct:** The design should be correct. The goal of the design process is not simply to produce a design of a system. But, the goal is to find the best possible design from the different alternative options.
   A design is correct if a system is built precisely according to design satisfies the requirements.

2. **Simple and Understandable:** It is the most important criteria for the software systems. The design of a system is one of the most important factors affecting the maintainability of the system.
   During maintenance, first necessary step that a maintainer has to understand the system to be maintained. A simple and

understandable design will make the job easier to the maintainer. Simplicity and understandability helps a lot in maintenance of software.

3. **Complete:** The design should be complete. It means that all the different components must be specified in the design. They include data structure, modules, external interface, interconnection between modules, etc. A design is complete if it implements all the specifications.

4. **Verifiable:** A design is verifiable if we easily verify whether design is correct or not. It is concerned with how easily the correctness of the design can be verified.

5. **Consistent:** It requires that there are no inconsistencies in the design.

6. **Efficiency:** It is concerned with the proper use of resources by the system. In computer system, the resources that are most often considered for efficiency are processor time and memory.

   Efficiency of any system is concerned with proper use of scarce resources. An efficient system is on that consumes less processor time and requires less memory.

7. **Traceable:** It requires that design of a system must be traceable in the requirement. It helps for design verification. A Design is traceable if all design elements can be traced to some requirements.

Input / Output / Exit Criteria of Design Phase
Input: SRS
Output: System Design & Detailed Design
Exit Criteria: Design which is verified & approved for quality.

2.0 DESIGN PRINCIPLES / CONCEPTS

Design Principles are guidelines that may be followed to create a good design of the system. The basic design principles for a designing a large system are:

1. Problem Partitioning
2. Abstraction
3. Top-down and Bottom-up strategy
4. Modularity

1. Problem Partitioning

It is possible to solve an entire problem at a first attempt for a small problem. But, it is not true for a large problem. When solving a small problem entire problem can be tackled at once. For solving larger problems, the basic principle is time tested principle of "Divide and Conquer".

It means that divide the system in to smaller pieces so that it can be conquered separately. A large problem can be easily solved using **"Divide and conquer"** principle.
This principle means that divide into smaller pieces, so that each piece can be conquered separately. Therefore, in a design of a system, the problem can be divided into manageably small pieces. Each of the pieces can be solved separately.

The reason behind solving problem dividing in pieces is that cost of solving entire problem is more than the sum of cost of solving all the pieces. Different pieces are not independent as they together form the entire system. It is at this point that no further partitioning is required. If a piece can be modified separately it is called independent of other pieces.

All the pieces work together to form a whole system. All of the pieces cannot be entirely **independent** of each other. The different pieces have to cooperate and communicate to solve the larger problem.

This communication adds the complexity, which arises due to partitioning together with cost of added complexity may become more than the savings achieved by partitioning. This **problem of complexity** arises due to a partitioning and it is not existing in our original problem.

A care should be taken, when we portioning a large problem. A design has to make the judgment about when to stop the portioning.

All the modules work together to form a whole system. All of the modules cann**ot** be entirely **independent** of each other. But, the design process should support as much independence as possible between modules. Dependence between modules is the one of the most reason for a high maintenance cost.

Problem partitioning is essential for solving a complex problem. Proper partitioning will make things easier to maintain by making design easier to understand. It is also helpful for verification of a design.

2. Abstraction

Abstraction means to consider what a component does, without considering how it does so. It is a tool that permits a designer to consider a component at an abstract level, without worrying about the details of the implementation of the component.

Abstraction of component describes external behavior of that component without bothering with internal details that produce the behavior.

It is the required part of the design process.

It is also essential for problem partitioning. Partitioning is a process to determine the component of a system. In partitioning we divide system in to smaller components.

To decide how a component interacts with other components, the designer has to know the external behavior of other components. For this he will concentrate on one component and try to understand the abstraction of other components.

The components of a system are not isolated from each other, but they are interacting with each other. It is responsibility of designer to specify the interaction between components.

An abstraction is used for existing component as well as new components.

Abstraction of existing components plays an important role in maintenance phase. For maintenance first step is to understand existing system for which we need to identify the abstractions of components. Using these abstractions working of entire system can be understood.
The basic goal of system design is to specify the modules in a system and their abstraction.

Once the different modules are specified, the designer can concentrate on one module at a time. The modules are implemented in so that abstract specifications of each module are satisfied.

There are two common abstraction mechanisms for software systems:

1) **Functional abstraction**
2) **Data abstraction**

1) **In functional abstraction,** the overall transformation function

for the function is partitioned into smaller functions that comprise system function. For example, a module to compute the log of a value can be represented by the function log.

2) **In data abstraction,** an operation is specified that are performed on the objects and data is not simply treated as objects, but is treated as objects with some predefined operations on them. The operations defined on data object are only operations that can be performed on those objects. From outside internals of the objects are hidden only the operations on the objects are visible. Data is an object and has some predefined operations can be performed on that objects.
Functional abstraction forms the basis of the structural design methodology, while data abstraction forms the basis of the object-oriented design methodology.

3. Top-down and Bottom-up strategy

A system is consisting of a hierarchy of a components. The highest level of a component is the total system. The lowest level of a component is the primitive component of a system.

There are two different approaches:

1) Top-down approach
2) Bottom-up approach

1) Top-down approach

The **top-down approach** starts from the highest level components of the hierarchy and proceed through to lower level.

A top-down design method is in the form of **"stepwise refinement".** Starting from an abstract design, in each step the design is refined to a more concrete level, until we reach a level where no more refinement is needed and the design can be implemented directly.

The top-down approach has been preferred by many researchers and has been found to be extremely useful for design.

A top-down approach is more suitable, if the specifications of the system are clearly known and the system development from the scratch. Many design methodologies are based on top-down approach.

2) Bottom-up approach

The **bottom-up approach** starts from the lowest level components of the hierarchy and proceed through to higher level.

A bottom-up design method is in the form of **"layers of abstraction".** Starting from the very bottom, operations that provide a layer of abstraction are implemented. The operations of these layers are used to implement more powerful operations until the stage is reached where the operations supported by the layer are those desired by the system.

A bottom-up approach is more suitable, if a system is to be built from the existing system. Pure top-down and pure bottom-up approaches are often not practical.

4. Modularity

Partitioning a system into modules is useful only if the modules are solvable separately.

A system is considered **modular** if it consists of discreet components so that each component can be implemented separately and change to one component has minimal impact on other component.

Modularity is a desirable property in a system. It helps in:

- **System debugging** as the system problem is isolated to a component,

- **System repair** as changing a part of the system is easy as it few other parts and
- **System building** as a modular system can be easily built by putting its modules together.

Modularity is where abstraction and partitioning come together. Modularity is a basic objective while partitioning and abstraction can be view as a concept.

3.0 MODULE LEVEL CONCEPT

A **module** is a logically separable part of the system. It is program unit that is discreet and identifiable with respect to compiling and loading. It can be a macro, a function, a procedure, a process, or a package.

Characteristics of modules

1. Module contains instructions, processing logic and data structures.
2. Modules can be separately compiled.
3. Module segments can be used by invoking and amend some parameters.
4. Modules can use other modules.

Modularization allows the designer:

- to decompose a system in to functional units,
- to impose hierarchical ordering on function usage,
- to implement data abstraction and
- develop independently useful system.

To produce modular designs, some criteria must be used to select modules so that they are solvable and modifiable. There are many criteria that can be used to guide the modularization criteria includes:

1. Conventional criterion

2. Information hiding criterion
3. Data abstraction criterion
4. Coupling & cohesion criterion
5. Problem Modeling criterion

In system using functional abstraction, **coupling and cohesion** are two criteria for selecting modules.

COUPLING

Two modules are considered independent if one can function completely without the presence of other. If two modules are independent, then they are solvable and modifiable separately.

But, it is not true because all the modules in a system **cannot** be independent of each other. All the modules must interact with each other to produce the external behavior of a system.

In short, all the modules work together to form a system. Therefore, they cannot be independent to each other.
If there is more connection between the modules, then the more knowledge about one module is required to understand and solve the other modules. It is easy to understand and solve the module, when there is a less and simple connection between modules.

The coupling is the concept of "**how strongly**" different modules are interconnected.

Definition: Coupling between modules is the strength of interconnection between modules or a measure of interdependence among modules.

Coupling can be of two types:

1. Highly coupled
2. Loosely coupled

A measure of interdependence among modules:

- Highly coupled modules are joined by strong interconnections.
- Loosely coupled modules are joined by weaker interconnections.

A high interaction between the modules is known as highly coupled while a low interaction between the modules is known as loosely coupled.

There is no interconnection between modules for independent modules.

There is interconnection between modules for dependent modules. To solve and modify a module separately, it is required that less interconnection between modules i.e. loosely coupled.

Coupling is an abstract concept and there is no formula can be given to determine the coupling between modules.
But there are some factors that can affect the coupling between modules. The major factors influencing coupling between modules are the type of connection between modules, the complexity and obscurity of interface between modules.

They are shown in the following table.

	Interface Complexity	**Type of Connection**	**Type of Communication**
Low	Simple obvious	To module by name	Data Control
High	Complicated obscure	To internal elements	Hybrid

Coupling increases with the complexity and obscurity of the interface between modules. To keep coupling low we minimize the number of interfaces per module and complexity of each interface.

Connections established by referring to other module names are more loosely coupled than the connection established by referring to the internal elements of other modules. In latter case, the entire content of referenced module may have to be taken in to account when updating the module that refers to it.

Complexity of the interface is another factor affecting coupling. The higher degree of coupling is required from a complex interface.

The type of information flow along the interface is the third major factor affecting coupling.

The more complex each interface is, the higher will be the degree of coupling. i.e., Complexity of entry interface of procedure depends on the number of times being passed as parameters and on the complexity of items.
Some level of complexity of interfaces is required to support the communication needed between the modules. However, more than this minimum is used.

For e.g. if a field record is needed by a procedure often the entire record is passed rather than just passing the field. This increases coupling unnecessary.

There are two types of information that can flow from one module to another.

1. **Control Information:** Passing and receiving control information means that the action of module will depend on this control information which makes it more difficult to understand.
2. **Data Information:** Data information means that a module passes as input some data to another module and gets in return some data as output.

Interfaces with data information only result in lowest degree of coupling.

COHESION

The cohesion is the concept of "**how closely**" the elements of a module are related to each other.

Definition: Cohesion of a module is a "**how tightly**" bound the internal elements of the modules are to one another.
Cohesion and coupling are related to each other. Normally, the greater the cohesion in the system the lower the coupling between modules.

There are several **levels of Cohesion**. They are:

1. Coincidental
2. Logical
3. Temporal
4. Procedural
5. Communicational
6. Sequential
7. Functional

1. Coincidental

Coincidental is the lowest level of cohesion. Coincidental cohesion occurs when there is no meaningful relationship among the elements of a module. It can occur if an existing program is modularized by dividing it into pieces and making different pieces modules.

For example, if a module is created to save duplicate code by combining some part of code that occurs at many different places, that module is likely to have coincidental cohesion.

In this situation, the statements in the module have no relationship with each other. If one of the module using the code needs to be modified then this modification includes common code, it is likely that other modules using the code do not want the code to be modified and may behave incorrectly.

The modules using such common modules are therefore not modifiable separately and have strong interconnections between them.

2. Logical

A module has **Logical** cohesion if there is some logical relationship between the elements of a module and the elements perform functions that fall in the same logical class. **For example**, a module that performs all the inputs or all the outputs.

3. Temporal

Temporal cohesion is the same as logical cohesion except that the elements are also related in time and are executed together. **For example,** a module that performs activities likes initialization, clean-up and termination are temporal cohesion.

4. Procedural

In **procedural** cohesion, a module contains elements that have a common procedural unit. **For example,** a loop or a sequence of decision statement in a module may be combined to form a separate module.

5. Communicational

A module with **communication** cohesion has elements that are related by a reference to the same input or output data. That is, in this cohesion, all the elements are work together because they operate on the same input or output data. An **example** of this could be a module to "print and punch record".

6. Sequential

A **sequential** cohesion exists when output of one forms to input to another in a module. A sequentially bound module may contain several functions or parts of different function.

7. Functional

Functional is the highest level of cohesion. It is the **strongest** cohesion. In this cohesion, all the elements of the module are related to performing a single function. **For example, a** module likes "Compute square root" and "sort the array".

4.0 STRUCTURED DESIGN METHODOLOGY (SDM)

SDM views every software system as having inputs that are converted to desired outputs.

Here entire system is viewed as transformation function that transforms the given inputs to outputs.

So designing a software system is called designing transformation function.

SDM is function oriented and heavily based on functional abstractions and functional decomposition.

The **aim of SDM** is to design a system so that the programs implementing the design would have a nice hierarchical structure.

A module with subordinates does not actually perform much computation. The actual computation is performed by its subordinates, and the module largely coordinated data flow between subordinates to get computation done.

The subordinates in turn can get bulk of their work done by their subordinates until their atomic modules have no subordinates are reached.

Factoring is the process of decomposing a module so that the bulk of its work done by its subordinates.

A system is said to be completely factored if all the actual processing is accomplished by bottom level atomic modules if non – atomic modules largely perform the jobs of control and coordination.

There are four major steps:

1. Restate problem as a DFD.
2. Identify the most abstract input and output data elements.
3. First-Level Factoring.
4. Factoring of input, output and transform branches

5.0 FUNCTIONAL V/S OBJECT – ORIENTED APPROACH

	Function Oriented Design approach		Object Oriented Design approach
1.	The design consists of module definitions with each module supporting a functional abstraction.	1.	In object-oriented approach the modules represent data abstraction.
2.	A system is viewed as a transformation function transforming input to output	2.	A system is viewed as a set of objects providing some services.
3.	It identifies hierarchy of modules	3.	It identifies objects and relationships between them
4.	Modules cannot be reused	4.	Modules can be reused
5.	It does not allow building object libraries	5.	It allows building object libraries.

6.0 DESIGN SPECIFICATION:

A conceptual design of the system can be produced in terms of structure chart here the functionality of module is communicated by name of the box and interface is indicated by data items labeling arrows.

This is all right while the designer is designing but inadequate while communicating the design to other programmer for future reference.

To avoid this design needs to be formally specified. The design specification is then communicated in the design document which is used as design reference for the design phase for later activities like detailed design and implementation.

It should contain all the information relevant to future phases.
A design specification should contain:

1. Problem specification
2. Major data structures
3. Modules and their specifications
4. Design decisions

1. Problem specification

Problem specification or restatement is the first step in the design process.

The requirement document specifies the problem in terms of the problem domain. And it is not suitable for design. The requirements are translated into problem specification. This problem specification can be well suited for designing a system.

2. Major Data structures

During the system design the major data structures for the software are identified and their formal definition is given modules and their specifications.

A design of a system has no meaning without specifying a data structures.

3. Modules and their specifications

Module specification is the major part of the system design specification.

All modules in the system should be identified when the system design is complete and these modules should be specified in the document.

To specify a module, the design document must specify:

- The interface of the module
- The abstract behavior of the module
- All other modules used by the module being specified.

4. Design decisions

All the major decision related to design made by the designer itself. A designer has to explain design decision explicitly. This makes a design more visible and will help in understanding the design.

VERIFICATION

Once the design process is over the output of process that means design of the system must be verified before processing with the activities of next phase.

Design Reviews

The purpose of design review is to ensure that the design satisfies the requirements and is of good quality. Errors in design process are reflected in the rest of the phases and the final system.

The cost of the solving errors increases as we move on to subsequent phases. Therefore it is best if design errors are detected early. The aim of the design reviews is to detect errors.

In system design review process, a group of people get together to discuss design with the aim of revealing design errors.

The review group includes:

- a member of the system design team and detailed design team,
- the author of requirements document,

- the author for maintaining the design document and
- an independent software quality engineer.

Here each member studies the design before the meeting, and with the aid of checklist marks items that reviewer feels are incorrect or need clarification. The member asks questions and the chief designer tries to explain the situation. During this discussion design error are revealed.
The psychological frame of mind should be healthy and one should not try to attack the ability of the designer. The meeting ends with a list of action items which are later acted on by design team.

Generally, design errors occur due to omission or misinterpretation of specified requirements or due to ambiguities in requirements.

Automated Cross Checking

In this method we check whether the design is internally consistent or not. For example, those modules used in a module that is defined in the system must also be defined in the design.

One should check whether the interface of module is consistent or with the way in which other modules use it.

Other internal consistency issues can be consistent use of data structures and whether data usage is consistent with declaration.

These consistency issues can be checked during design review and is usually the place where they are checked if no automated help is available. If the design is expressed in a language designed for machine processing, most consistency checking can be automated.

7.0 DETAIL DESIGN

A **detailed design** is second step of a design phase. It specifies internal logic of each module of a system.

The input of a detailed design is system design document. The output of a detailed design is detailed design document.

During this phase, details of data structures and algorithm design of each module is specified. The logic of a module is specified in high-level language. It is also known as **logic design**.

Importance of detailed design

In system design we concentrate on the module in a system and how they interact with each other. In a design document, a more detailed specification is given by explaining in natural language what a module is supposed to do.

These non-formal methods of specification can lead to problems during coding, particularly if the coder is different from person from the designer.

The reason is that the correct implementation of the module depends on the coder interpreting the non-formal specifications of modules in precisely the manner the designer intended.

Even if the designer and coder are same person, problems can occur as the design may take long time and the designer may not remember precisely what the module is supposed to do.

Formal specifications can ensure that the specifications are precise and not open to multiple interpretations. Hence, we need to formally specify a module.

MODULE SPECIFICATION

The detailed specification about what a module does is called module specification.

The desirable **properties / characteristics** of a Module should have:

1. **Complete:** Specifications must specify entire behavior of the module.
2. **Unambiguous:** Specifications must be clearly identified.
3. **Understandable:** It should be written in such a language that it could be understood.
4. **Consistent**
5. **Modifiable**
6. **Traceable**
7. **Implementations independent:** It must specify only external behavior of the system and should not specify internal details of the module. It must not specify any method for implementation.
8. **Operational Specifications:** It must specify only external behavior of the system and should not specify internal details of the module. It must not specify any method for implementation.

Formal method of module specification can ensure that the specifications are precise and do not have multiple interpretations. The formal specification methods for the modules are:

1. Specifying functional modules (Functional abstraction)
2. Specifying Classes (Data abstraction)

FUNCTIONAL MODULE SPECIFICATION

The view of a functional module is like a **Black box**.

In functional module specification, a module is treated as a black box that takes an input and produces some output such that the output has specified relationship with the inputs.

A module can operate only on inputs that satisfy some constraints. The constraints may be on type of input and the range of inputs.

For example, a function that finds a square root of a number may be design to operate only on the positive real numbers.

In short, a module has to produce the correct output on a correct input.

One method for specifying modules was proposed by Hoare, based on **pre-condition and post-condition**.

In this method, the input constraints of a module are specified by a logical assertion on the input state is known as **pre-condition.**

In this method, the output from a module are specified by a logical assertion on the output state is known as **post-condition**.

There is no relationship is explicitly specified between the input and the output.

Validity of the output is specified entirely by the post-condition assertion.

Example: Consider a module 'SORT' to be written to sort a list L of integer in ascending order.

The pre-condition and post-condition of this module are:
Pre-condition: Non-null L

Post-condition: For all i, 1 <= i < size(L), L[i] <= L[i+1]

These specifications are not complete. The post-condition only state that the final state of the list L should be such that the elements are in ascending order. But, it does not state that final state of the list L should contain the same elements as the initial list.

The pre-condition and post-condition of this module are rewritten as: Pre-condition: Non-null L

Post-condition: For all i, 1 <= i < size(L'), L'[i] <= L'[i+1] and

L' = permutation(L)

This specification states that elements of the final list are a permutation of elements of the initial list. Now, this specification is complete.

The module checks if the input satisfies the desired constraint. If it does not satisfy then it is treated as an **exception condition** and some special code is executed.

DATA ABSTRACTION

Data abstraction is one of the most important language concepts of recent times.

It is consist of a group of related operations that act on a particular object with the constraint. The behavior of the object can only be observed by applying operations on an object.

It is useful for information hiding. It provides a high level of abstraction.

A data type in a language that supports data abstraction is known as Abstract Data Types (ADT).

Many languages support Abstract Data Types (ADT) such as SIMULA, ADA, C, C++, JAVA.

Various specification techniques have been used for specifying Abstract Data Types. The

Axiomatic Specification Technique is one of them.

Axiomatic Specification Technique:

It is a technique has been used to specify Abstract Data Type. In this technique, the behavior of each operation is not directly specified. But, it is specified by axioms.

Example: Consider an example of a stack of an integer. We define a stack that has four operations:

- Create : To create a new stack
- Push : To push an element on a stack
- Pop : To pop the top element from the stack
- Top : To returns the element on top of the stack.

The axiomatic specification of a stack is as shown below:

stack [integer] (Declaring the operation)

declare

create() → stack ; (As a result gives stack)

push (stack , integer) → stack ;

pop (stack) → stack ; (Update stack with 1 element)

top (stack) → integer u undefined ; (Undefined means stack is empty)

var

s : stack ;

i : integer ;

for all

top (create ( )) = undefined ;

top (push (s, i)) = i ; (i th element is pushed)

pop (create ( )) = create() ;

pop (push (s, i)) = s ; (we have inserted element and taking it out so it will result into updated stack)

end

The above axiomatic specification language has two major components. They are:

- Syntactic specifications
- Semantic specifications

The **syntactic specifications** provide the syntactic and type checking information such as variable names, variable types and domain and range of operations. The syntactic part of the specification has three components:

1. The header
2. Operation declaration
3. Variable declaration

The header : The header specifies the name of the data types and parameters.

Operation declaration: All the operations of the ADT must have declared in the operation declaration part. For an operation, the declaration has to specify the type of the input parameters and the result. The type can be ADT itself, any of the type parameters of the ADT or a standard type like Boolean or integer.

Variable declaration: The variables of specified types can be declared here. It can be used in semantic specification.

The **semantic specifications** define the meaning of the operations by stating relationship of the operations among each other.

The semantics of the operations in the axiomatic specification technique are specified in the semantic part by axioms. Axiom attaches meaning to operations by specifying the relationship between operations.

The following constructs are allowed for writing axioms:

- Free variables
- If-then-else

- Recursion
- Boolean expression

8.0 VERIFICATION OF DETAILED DESIGN

The verification of detailed design is a process to verify that the detailed design meets the specifications given in the system design.

There are three verification techniques used to verify the detailed design documents. They are:

1. Design Walkthroughs
2. Critical Design Review
3. Consistency Checkers

1. Design Walkthroughs

It is a manual method of verification.

The group of design walkthrough is usually small. This group contains the designer, group leader and other members concern with detailed design.

It is done in an informal meeting called by the designer or the leader of the designers group.

In a meeting, the designer explains the logic step by step. The members of the group ask the questions, point out the possible errors or ask for clarification about doubts.

A beneficial side effect of the walkthrough is that in the process of explaining the design in detail, the designer himself can cover some of the errors.

Due to its informal nature, this technique is NOT as effective as the critical design review.

2. Critical Design Review

The **purpose** of critical design review is to ensure that the detailed design satisfies the specifications given in system design.

This review is similar to other review in which a group of people work together to discuss the design. During the discussion, they find the design errors or undesirable properties of a design.

The **review group** includes the author of the detailed design, a member of a system design team, the programmer responsible for coding and independent software quality engineer.

Each member studies the design before meeting. The member of a team comes up with a **checklist.**

A sample check list contains:

- Does each of the modules in the system design exist in detailed design?
- Are there analyses to demonstrate that the performance requirements can be met?
- Are all the assumptions explicitly stated and are they acceptable?
- Have the exceptional conditions been handling?
- Are all the data formats consistent with the system design?
- Is each statement specified in natural language easily codable?
- Are the loop termination conditions properly specified?
- Is nesting proper?
- Is the module logic too complex?
- Are the modules highly cohesive?

At the meeting, the members of a team ask questions and the designers tries to explain the situation. During the discussion design errors are discovered.

It should be kept in mind that the aim of the meeting is to uncover the design errors, not try to fix them. Fixing of an error is done later.

The meeting should **end** with a list of action should be taken by the designer at later stage.

3. Consistency Checkers

The design is specified in PDL or some other formally defined design language.

The consistency checkers is used to detect the design defects.

Consistency checkers are essentially compilers that take as input the design specified in a design language like PDL and detect some design defects.

A consistency checker can ensure that any modules invoked or used by a given module actually exist in the design and that the interface used by the caller is consistent with the interface definition of the called module. It can also check if the used global data items are indeed defined globally in the design.

CHAPTER - 4

1.0 CODING

Coding is the process to translate design into the code of required programming language.

Input: It contains two documents:

1. System design documents and
2. Detailed design documents.

Output: The unit tested code of all modules.

Description:

This phase is important and it affects the testing and maintenance. If the code is well written, it can reduce the cost of testing and maintenance. So actually the goal of this phase should be to reduce cost of testing and maintenance. This means that the programs should not be written so that they are easy to write, but it should be easy to read and understand.

The coding should be a Structure Programming. The goal of structured programming is that control flow is linear in the program. It means that the code should be written in a sequence of steps and also executed in the sequence.

Top-Down and Bottom-Up approach for coding

Once the design is already over, we can implement that design in terms of coding by top-down or bottom-up approach.

1) Top down means that we start to implement the main modules first and then start the coding of the sub modules and so on.
It is used in system where detail design in not enough. When we develop a prototype, some of the design decisions have to be made during development. In such a case top-down approach is most preferable technique.

2) An implementation of bottom-up coding is the reverse process of a top-down implementation.
Bottom-up means that we start from the modules at lowest level and then code for the higher modules. It is used for layered systems like O.S. and network software.
This approach is not concerned with the design approach i.e. design may use bottom up approach, but we can use top down to make coding of it. However, the same approach with design is preferred.

2.0 STRUCTURED PROGRAMMING

Generally, the structured programming means" go to less programming ". But we can also write structured programs with go-to.

A program has static and dynamic structure:

- **Static** means the lines of code in which it is written.
- **Dynamic** means the sequence of lines which will be executed in program when it runs.

The goal of structured programming is that static & dynamic structures are same. All the programs contain the branching or repetition. So, the statements should be single-entry-single-exit, i.e. it should start from one point and end at one point.

Examples are as below.

Selection	: if B then S1 else S2
Iteration	: While B do S or Repeat until B
Sequencing	: S1:S2:S3

Thus, the conclusion is that programming should be such that it uses these types of statements.

INFORMATION HIDING

When software is developed to solve a problem, the software uses some data structures to capture the information in the problem domain.

In general, only certain operations are performed on some information.

For example, a ledger in an accountant's office has some very defined uses: debit, credit, check the current balance, etc. An operation where all debits are multiplied together and then divided by the sum of all credits is typically not performed.

So, any information in the problem domain typically has a small number of defined operations performed on it.

When the information is represented as data structures, the same principle should be applied, and only some defined operations should be performed on the data structures. This, essentially, is the principle of information hiding.

The information captured in the data structures should be hidden from the rest of the system, and only the access functions on the data structures that represent the operations performed on the information should be visible.

In other words, when the information is captured in data structures and then on the data structures that represent some information, for each operation on the information an access function should be provided.

And as the rest of the system in the problem domain only performs these defined operations on the information, the rest of the modules in the software should only use these access functions to access and manipulate the data structures.

Information hiding can reduce the coupling between modules and make the system more maintainable.

Information hiding is also an effective tool for managing the complexity of developing software—by using information hiding we have separated the concern of managing the data from the concern of using the data to produce some desired results.

Many of the older languages, like Pascal, C, and FORTRAN, do not provide mechanisms to support data abstraction. With such languages, information hiding can be supported only by a disciplined use of the language.

That is, the access restrictions will have to be imposed by the programmers; the language does not provide them. Most modern OO languages provide linguistic mechanisms to implement information hiding.

PROGRAMMING STYLE (Do's and Don'ts of Coding or Programming)

It is not possible to give the list of rules for a good programming. But there are certain general rules which can be used, which are as below:

1. **Names:** The programmers do not consider important to name the module or variable. The names should be given so that it is closely related to its use.

   The three things should be avoided:
   - giving cryptic (short) names,
   - unrelated names and
   - same name for multipurpose.

2. **Control constructs:** Generally single-entry-single-exit types of constructs should be used.

3. **GoTos:** It should be used only if there is no other choice. The forward gotos are more acceptable than the backward. The use of goto for exiting and error handling is also acceptable.

4. **Information hiding:** As above, the data abstraction should be used if possible.

5. **User Defined types:** The languages support this type of data. It should be used whenever required.

6. **Nesting:** The deep nesting should be avoided. For e.g. if there are statements as:

```
If Cl then S1
          else if C2 then S2
                    else if C3 then S3
                              else if C4 then S4
```

These types of statements are difficult to understand. So, we can convert it to:

```
If Cl then S1
If C2 then S2
If C3 then S3
If C4 then S4
```

But we get little inefficiency in such statements with advantage of easy reading.

7. **Module size:** The programmer should check that the modules are not too small (less than 5 lines) or even too large (more than 100 lines). This is because it has the effect on the coupling and cohesion.

8. **Module Interface:** The modules should not have complex interface. For eg. If module has more than 5 parameters dependent on other module, it should be divided into two separate modules.

9. **Program Layout:** It means that programs should be such that it is easy to understand.

10. **Side Effects:** If module is changing some global variable, such side effects should be documented.

11. **Robustness:** A program is robust if it does something even when the exception conditions occur. Thus, program should have validation, so that it does not give problems like "crash" or "core dump", but should give meaningful message.

INTERNAL DOCUMENTATION

The output of coding phase is the code itself. So, the proper comments should be written in the programs for the better understanding.

The comment should specify what the code is doing and not how it is doing. It is not required for comments for each line. Generally, the comment for whole module is enough.

This must include:

- Functionality of module
- Parameters and their purpose
- Assumptions of input
- Global variables used or modified

In addition to above, other information can also be included, which is decided as standard for the system, like- Author name, Compilation date, Modification date etc.

VERIFICATION OF CODE

The goal of verification is to check that the code implemented is according to the design.

Verification of program can be static or dynamic.

- **Static** method do not involve actual program execution on actual data but it focuses only the conceptual execution,

- In **dynamic** method, the program is executed with some test data and the output of program is checked to find the errors.

Common forms of static techniques of program verification are: Code reading, Static analysis, Code reviews and walkthrough, and Symbolic execution.

Code reading

Code reading technique is used for verification. In this process, the programmers read the code and find the difference between the design specification and actual implementation.

This process is best if the code is read from inside-out, i.e. we use the abstraction. It is very useful method, and find the errors which are not found in the testing.

This method will result into well- structured program and so it is also called **desk review**.

TESTING FUNDAMENTALS

Testing: Testing is evaluation of the software against requirements gathered from users and system specifications. Testing is conducted at the phase level in software development life cycle or at module level in program code.

Error: Error can be interpreted with two ways. It is the difference between computed value and theoretically correct value. In terms of software, it is a difference between the actual output and the ideal output required. That results in software containing defects.

Errors are of two types: **(1) Programming Error (2) User Error.**

Fault: Fault is a condition that causes a system to fail in performing its required function. A fault is the basic reason for software malfunction and is synonymous with the commonly used term bug.

Failure: Failure is the inability of a system or component to perform a required function according to its specifications. A software failure occurs if the behavior of the software is different from the specified behavior.

Failures may be caused due to functional or performance reasons. A failure is produced only when there is a fault in the system. However, presence of a fault does not guarantee a failure.

Reliability: Reliability is the probability that software will fail for the specified time and specified condition.

Justify "When to stop testing" is one of the hard issues.

A failure is produced only when there is a fault in the system. However, presence of a fault does not guarantee a failure.

In testing, we observe the behavior of the system or component and try to determine whether there is failure or not.

Because of this fundamental reliance on external behavior observation, testing can only reveal the presence of fault, but by not observing a failure during our observation interval we cannot claim that there are no faults in the system.

Thus, it becomes hard to decide for how long we should test a system without observing any failures before deciding to stop.

Justify "Testing is an expensive method for identification of faults, compared to methods that directly observe faults"

During testing process only failures are observed and presence of faults is reduced. The actual faults are identified by separate activities, commonly referred to as "debugging"

Thus after testing, expensive task of debugging has to be performed. Hence, then we can say above statement.

TOP DOWN AND BOTTOM UP APPROACHES FOR TESTING

In top-down approach for testing, we start with the top of the hierarchy and then add all the modules that it calls.

Here we require stub- which are dummy routines that simulate the module, which is to be tested.

In **bottom-up**, we start with the bottom and then they are combined with higher level modules. To find these modules, we require drivers to call the parent modules.

Top down is useful if major errors are at top and bottom up is useful if major errors are at bottom. However, the method used depends on the method of software development.

Thus, if software is developed with top- down, we should use top-down for testing also. Following table distinguish between both approaches.

	Top Down	**Bottom Up**
1	Top-down integration testing is an integration testing technique used in order to simulate the behavior of the lower-level modules that are not yet integrated. Stubs are the modules that act as temporary replacement for a called module and give the	Bottom up integration testing also uses test drivers to drive and pass appropriate data to the lower level modules. As and when the code for the other module gets ready, these drivers are replaced with the actual module.

	same output as that of the actual product.	
2	Advantageous if major flaws occur toward the top of the program.	Advantageous if major flaws occur towards the bottom of the program.
3	In top down approach, main () function is written first and all sub functions are called from main function. Then, sub functions are written based on the requirement.	In bottom up approach, code is developed for modules and then these modules are integrated with main () function.
4	Top down approach begins with high level design and ends with low level design or development.	Bottom up approach begins with low level design or development and ends with high level design.
5	Structure/procedure-oriented programming languages like C programming language follows top down approach.	Object oriented programming languages like C++ and Java programming language follows bottom up approach.
6	Top-down approach analyzes risk by aggregating the impact of internal operational failures	Bottom-up approach analyzes the risks in individual process using models
7	Top-down approach doesn't differentiate between high frequency low severity and low frequency high severity events	Bottom-up approach does differentiate between high frequency low severity and low frequency high severity events
8	Top-down approach is simple and not data intensive	Bottom-up approach is complex as well as very data intensive
9	Top-down approaches are	Bottom-up approaches are

	backward-looking	forward-looking
10	Top Down --> BIG SYSTEM to smaller components	Bottom Up --> smaller components to BIG SYSTEM

LEVELS OF TESTING

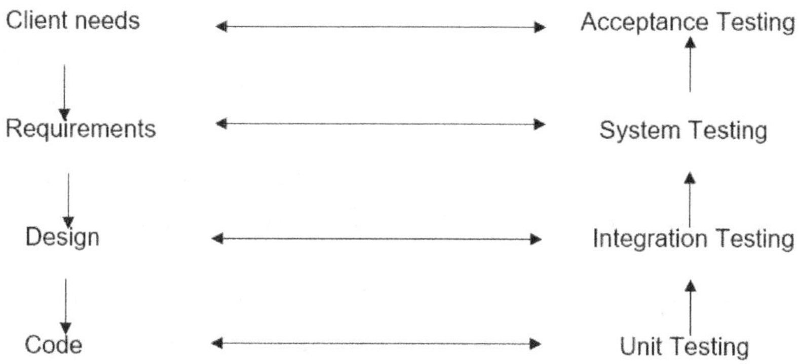

The testing process is conducted after each phase, as there are chances of errors in all of them. For this reason, there are different levels of testing used, as above.

The basic levels of testing are:

1. Unit testing
2. Integration testing
3. System testing and
4. Acceptance testing

1. Unit Testing

It is a process to test each of the modules. In this, different modules are tested against the specifications produced during design for the modules.

Unit testing is essentially for verification of the code produced during the coding phase and hence the goal is to test the internal logic of the modules. It is done by the programmer of the module.

2. Integration Testing

It is a process to test the system after combining all the modules. In this, many unit-tested modules are combined unto subsystem, which are then tested.

The goal of integration is to see the modules are integrated properly. Here, the goal is to test that the interface between module is proper or not. This level is thus related to the testing of a design.

3. System Testing

The system testing is a testing to check that system is working according to requirements. The entire software is tested.

Here, the goal is to see that the software made is according to the requirement document. Here, all the validations are performed.

4. Acceptance Testing

The acceptance testing is a testing to check the working of software according to actual data. The realistic data is given by the client to check the software working satisfactorily.
Here, the goal is to check the external behavior of system, and not the internal logic.

Note: Another level is **regression testing**, which performed when some changes are made to an existing system.

As modifications have been to an existing system, testing has to be done to make sure that new features indeed work and has not had any side effects on older services.

For regression testing old test cases are again executed on the modified system and its output is compared with earlier output to make sure that system is working as before on these test cases.

FUNCTIONAL TESTING V/S STRUCTURAL TESTING

	Functional Testing	Structural Testing
1.	Functional Testing is also called "black box" (behavioral) testing.	Structural Testing is also called "white box" testing.
2.	Structure of program is not considered for testing.	Structure of program is considered for testing.
3.	It is concerned with testing functionality of program.	It is concerned with testing implementation of program.
4.	Compares program performance with the program specification	Does not test the software against the program specification
5.	Test cases are decided on the basis of specification of program or module because you do not have access to the program code and you do not have any knowledge of the internal structure of the program.	Test cases are decided on the basis of internal logic of program or module
6.	Criteria for testing are imprecise.	Criteria for testing is precise and formal.
7.	Testing is performed by the software testers.	Testing is performed by the software developers.

AUTOMATED TESTING TOOL

Automated Testing is a technique where Tester Writes Scripts by own and uses suitable Software or Automation Tool to test the software. It is Automation Process of a Manual Process. Allows to execute repetitive task without intervention of Manual Tester.

Automation testing means running the software programs that carry out the execution of test cases automatically and produce the test results without any human intervention.

It is one step ahead of the manual testing. It saves the human effort and time up to a great extent and it also leaves no or very less scope for errors in testing. Once ready, automated tests can be run any number of times to test the same application thus minimizing redundant manual work.

Automated testing differs from manual testing where a human being is responsible for single-handedly testing the functionality of the software in the way a user would.

Because automated testing is done through an automation tool, less time is needed in exploratory tests and more time is needed in maintaining test scripts while increasing overall test coverage.

The benefit of manual testing is that it allows a human mind to draw insights from a test that might otherwise be missed by an automated testing program.

Automated testing is well-suited for large projects; projects that require testing the same areas over and over; and projects that have already been through an initial manual testing process.

Automation Testing Process

- **Test Tool Selection:** There will be some criteria for Selection of tool. Majority of the criteria includes: Do we have skilled resource to allocate for automation tasks, Budget constraints, Do the tool satisfy our needs?
- **Define Scope of Automation:** This includes few basic points such as: Framework should support Automation Scripts, Less Maintenance must be there, High Return on Investment, Not much complex Test Cases
- **Planning, Design and Development:** For this we need to Install particular frameworks or libraries, start designing and developing the test cases such as NUnit, JUnit, QUnit or required Software Automation Tools

- **Test Execution:** Final Execution of test cases will take place in this phase and it depends on Language to Language such as for .NET, we'll be using NUnit, for Java, we'll be using JUnit, for JavaScript, we'll be using QUnit or Jasmine, etc.
- **Maintenance:** Creation of Reports generated after Tests and that should be documented so as to refer that in future for next iterations.

Popular Automation Tools

- Selenium
- QTP
- Sikuli
- Appium
- Jmeter
- TestNG
- Zephyr
- UFT (HPE Unified Functional Testing)

Advantages of Automation Testing

- Simplifies Test Case Execution
- Improves Reliability of Tests
- Increases amount of test coverage
- Minimizing Human Interaction
- Saves Time and Money
- Test Results are made public

Made in the USA
Monee, IL
04 November 2026

67379254R00056